Red Light Therapy

A Complete Guide to Red Light Treatment

Kathy Richards

Your Free Gift

As a way of thanking you for the purchase, I'd like to offer you a complimentary gift:

- **5 Pillar Life Transformation Checklist:** This short book is about life transformation, presented in bit size pieces for easy implementation. I believe that without such a checklist, you are likely to have a hard time implementing anything in this book and any other thing you set out to do religiously and sticking to it for the long haul. It doesn't matter whether your goals relate to weight loss, relationships, personal finance, investing, personal development, improving communication in your family, your overall health, finances, improving your sex life, resolving issues in your relationship, fighting PMS successfully, investing, running a successful business, traveling etc. With a checklist like this one, you can bet that anything you do will seem a lot easier to implement until the end. Therefore, even if you don't continue reading this book, at least read the one thing that will help you in every other aspect of your life. Grab your copy now by clicking/tapping here or simply enter http://bit.ly/2fantonfreebie into your browser. Your life will never be the same again (if you implement what's in this book), I promise.

PS: I'd like your feedback. If you are happy with this book, please leave a review on Amazon.

Introduction

Also called biostimulation, light box therapy, low-level light therapy (LLLT), or Photobiomodulation, Red Light Therapy or treatment has gained popularity.

Although the therapy is what we would consider "alternative," many are swearing by it and using it to enhance wellbeing, promote recovery after surgery, minimize the effects of aging (to reduce wrinkles), and for various other personal wellbeing ends including but not limited to improved hair growth, easing sore muscles, dry skin, winter depression, and even weight loss.

This complete Red Light Therapy guidebook looks at everything you need to know about using Red Light Therapy.

Among other things, we shall seek to:

1. Cultivate a deeper understanding of Red light therapy.

2. Determine whether it works and how it works,

3. The various benefits of using this therapy, and everything else you would need to know about this alternative therapy especially if your intent is to use the therapy to promote personal wellbeing.

Let's get to it:

Table of Content

Medical Disclaimer

The content of this book is for informational purposes only. It does not to any extent intend to substitute any professional medical advice, treatment or diagnosis. You should always seek the guidance or advice of your health care provider in case you have any questions regarding a serious medical condition.

A Beginner's Guide to Red Light Therapy

Red light therapy or treatment is *an alternative, heliotherapy treatment/therapy that exposes the skin to red light and near-infrared light at very specific wavelengths—630-880—as a way to treat skin-related issues such as scars, wounds, and wrinkles, and to enhance overall personal and cellular health and wellbeing."*

The reasoning and science behind using red Light therapy is not new. It dates back millennia, which is perhaps why to cultivate a deeper understanding of what Red Light Therapy is, we need to look at how we (humankind) have been using light—natural and artificial—for various means and ends including using it as a form of treatment.

A Bit Of History

As the name suggests, red light therapy is a type of *heliotherapy* or light therapy; heliotherapy is a type of therapy where the aim is to expose the skin to light at specific wavelengths. We have been using heliotherapy for thousands of years if not from the beginning of modern civilization.

For example:

Historical researchers believe that residents of Ancient Egypt, Rome, and Athens used different forms of light therapy—such as using the light of the sun to enhance overall wellbeing—and that as early as 1500 BCE, Ancient Indians were combining their knowledge of herbs and the light from the sun to treat various skin ailments such as wounds and scabs.

Modern day phototherapy, which is where red light therapy falls, owes a debt of gratitude to Neil Finsen, a physician of Faroese descent who many believe to be the first person to develop an artificial light source designed for the specific purpose of using the light for healing purposes. He used the light to treat skin infections such as lupus vulgaris.

Wikipedia notes that later—late 1800-early 1900s—Finsen fashioned *red lights* that he then used to treat lesions caused by smallpox—for which he received the 1903 Nobel prize in Medicine.

Since the early 1900s, heliotherapy has seen great advances. We have also used it for different means including using specially created red lights for food production—in the 1990s, NASA used such a technology to grow food in space—and healing as is the case with RLT and other forms of heliotherapies such as LED light treatment.

Most heliotherapies are very particular about the amount and nature of exposure, i.e. the wavelength of the light and the amount of time one remains exposed to the 'therapeutic' light. The reason for this is that most modern therapies under the light therapy (or heliotherapy) classification use specialized lights technologies such as lasers, full-spectrum lights, LED light, polychromatic polarize lights, or dichroic lamps. Although using these specialized lights on the skin is safe, in some instances, over exposure to some of these lights can be damaging to the skin or treatment area, which necessitates caution.

Now that you know what Red Light Therapy (RLT) is as well as how it came to develop, let's move a step further and discuss how the technology and science driving it works:

How Red Light Therapy (RLT) Works

Dictionary.com describes an electromagnetic spectrum as *"the range or number of frequencies or wavelengths electromagnetic radiation can extend."* If you are wondering why this is important, here is why:

During a red light therapy session, the idea is to expose a fair amount or a particular amount of skin to a specific wavelength of red light for a specific amount of time, normally 10-60 minutes. From this exposure, the light penetrates the skin where upon interacting with the cell nuclei, it offers therapeutic benefits.

Red light is part of the larger electromagnetic spectrum consisting of seven electromagnetic waves: *Gamma Rays, X-rays, Ultraviolet Waves, Visible Light Rays, Infrared Waves, Microwaves and Radio Waves.*

On this spectrum, along with other hues of colors such as blue, violet, orange, and yellow, red light falls under the visible light rays classification, which means the light is visible to the naked eye.

Each of the colors within the larger visible-light-rays classification has a specific wavelength or frequency; to measure these frequencies, we use nanometers (nm). All visible light rays—or the wavelength of all lights within the visible-light-rays category—have a wavelength range of 400-880nm.

The length of the frequency determines the light's ability to penetrate the skin, i.e. longer wavelengths penetrate the skin deeper. Because various red lights have a wavelength capacity of 630-880nm, they have the ability to penetrate deeper into the skin—sometimes as deep as 5-10 millimeters–, which means the light has the potential to reach all layers of the skin.

The human skin is the largest organ of the integumentary system, an important bodily system that acts as a protective barrier against diseases/viruses and that helps the body retain fluids, regulate body temperature, and remove bodily waste. This means the skin has millions if not billions of cells.

Cells are the powerhouses that drive you; within cells are mitochondria, the billions of cell stations that your body uses to create energy on a cellular level. For optimal cellular health, adenosine triphosphate (ATP) is the most important molecule. Without it, our cells would starve, and without the energy they need for cellular metabolism—this includes internal and external bodily functions that require energy such as walking, digestion, healing, breathing, repair, etc.— they would die off, and so would we.

When you shine a therapeutic grade red light on the skin, because the light has the ability to penetrate the skin deeper and therefore make contact with billions of cells, it positively affects the functionalities of the mitochondria— power generators of the cell–by triggering the production of more ATP.

When the red light triggers enhanced production of ATP, the resulting biochemical effect enhances mitochondrial and cellular functionality—efficient energy production—and because of enhanced ATP production, the cells have easy access to the energy they need to repair, rejuvenate, and function optimally.

To have a better understanding of the effect red light therapy has on cells and production of ATP, you need to understand that even without "shining a special red light" on your skin and therefore cells, all your cells need natural light to power the cellular processes that produce the energy your body needs in your everyday life. Without red light therapy, your cells use natural red light to power this process.

The idea behind red light therapy is to control when you expose the skin to a healing spectrum of red light, and to shine a specific wavelength of light on the skin with the intended aim being to enhance cellular activity by triggering a leaner and more efficient ATP production process.

The science behind this process is that the body has two ways to create ATP: anaerobic and aerobic, using, and without using oxygen respectively. Cellular respiration is the process through which this happens. In cellular respiration, the body uses the various means at its disposal such as the food and water we eat, and the air we breathe to create the energy we need to survive and operate optimally. Of the two processes, aerobic ATP generation is more effective and because it produces the most energy, it is the body's favorite way of creating ATP and energy.

Aerobic ATP generation has four primary stages: the first two are *glycolysis* and *pyruvate oxidation*. In glycolysis, our body uses available resources (air, water, food) to synthesize pyruvates. In the second stage, pyruvate oxidation, the body takes the pyruvate and oxidizes it into Acetyl CoA. The Acetyl CoA production process leads to the generation of the carbon dioxide compounds that become the aerobic energy used by 95% of all the cells in the body.

The third and fourth stages are *citric acid cycle* and *oxidative phosphorylation*. In the citric acid cycle stage, oxidation is the primary objective. In this stage, the body takes the Acetyl CoA created in the second stage and oxidizes it to create the coenzymes NADH and FADHs. In the fourth stage, electrons conduct the coenzymes NADH and FADHs into the electron transport chain (ETC) and from the process, create energy.

Using a red light machine routinely as prescribed ensures that the nitric oxide created from the ATP creation process does not become too oxidative and therefore harmful to cells. By stimulating mitochondria into creating more ATP, cells function better, heal and repair faster, and are better capable of staving off ailments, viruses, and the likes.

In addition to enhancing ATP production on a cellular level, exposing the skin to therapeutic red light and near infrared lights (NIL) at specific and special wavelengths has also shown to promote better cellular communication and signaling. When the cells communicate better, the results are improved protein synthesis and better cell functionality and enzyme activity.

Because of their effect on the production of nitric oxide within the cells, red and NIL lights enhance the production of antioxidants. Antioxidants are important because they protect the cells from oxidative and chronic stress and mutations that may impair cell health, circulation, and other important bodily functions including those connected with health and overall wellbeing.

Unlike ultraviolet waves, microwaves, and higher spectrums of gamma rays, exposure to Red light and near-infrared wavelengths is not harmful because treatments occur in a controlled environment such as a licensed spa or medical center, or even using a high-grade red light device at home. This means the chances of the light wavelengths generated by pure red light burning the skin are very dismal. This makes using Red light therapy safe-to-use for most people.

NOTE: With the above mentioned, for safety purposes, if you do decide to look into using red light therapy, it is best to consult a health professional before using this specific therapy—and any other light therapy in general—and to ensure treatment sessions occur in a safe and regulated environment (your home also counts).

Now that you know how red light therapy works, the other question that may be on your mind is whether the therapy actually works. The next section looks at this by looking at the various, scientifically proven, benefits of red light therapy and heliotherapy in general.

Does RLT Work: The Scientific Proof/Benefits

To answer the question, RLT works. As scientific research has shown and proven, using it has various benefits, many of which we shall discuss at length in this chapter.

To recap what we learned in the last sections, RLT works by shining on the skin a penetrating red, therapeutic light that stimulates cellular production of ATP, the molecule that ensures transmutation of resources into the cellular energy our bodies need and use to power various metabolic processes including those related to cellular healing and rejuvenation.

Scores of research studies have shown that when used safely and consistently, the resulting biochemical process caused by exposure to red light therapy has the following potential benefits:

RLT Benefits the Skin

Various studies have concluded that controlled use of safe red light is immensely beneficial for the skin.

RLT therapy enhances the overall health of the skin. As we discussed earlier, when the red light impacts skin cells where it stimulates the production of ATP, this enhanced production leads to overall cellular wellbeing, which effectively translates to a smoother, acne free and healthier skin, boosted collagen production which means a wrinkle free skin, and because of enhanced cell health, better healing abilities and less skin scaring.

Many professionals, dermatologists, celebrities, estheticians, and important publications such as Elle, CNN, and others have publicly noted the positive effects red light therapy has on the skin.

To enhance your understanding of how red light therapy leads to the enhanced wellbeing of your skin, here is how the process works.

How this works

Shining a controlled wavelength of red light is the essence of red light therapy. As we discussed earlier, because of the controlled aspect of the red light therapy of wavelengths 660-850 nm, exposing your skin to these wavelengths of light is safe because even without shining a red light on the cells, they have the ability to absorb from the environment these wavelengths of light and use it to power cellular functionality.

When you take more control of the process by which your cells receive the wavelength of pure red light they need to function effectively, the resultant, controlled or personally instigated stimulation of ATP production leads to better skin functionality (remember that the skin is an organ too, in fact, one of the largest and most important organs).

When the skin cells function better and when the cells can circulate well within the various skin layers and the Electron

Transport Chain, it leads to lower oxidation. Lower oxidation means an efficient cellular energy production process, better circulation, functionality, reduced inflammation of the skin, better protection from harmful rays, and overall improved skin appearance and health.

In regard to increasing cells circulation within the various skin layers, a <u>research study conducted by Maria Emília de Abreu Chaves et al</u>. noted that continued controlled use of red light therapy in any form—even a handheld red light therapy device used at home—increases circulation by ensuring that the cells function effectively and the tissues receive the energy, nutritional, and oxygen support they need to function optimally.

Moreover, because the process also reduces oxidative stress on the cells, it enhances the skin's ability to detoxify itself—this means a lesser need for 'skin detoxes' and products—and to heal wounds and scar tissues better and much faster.

RLT is beneficial for the skin in a many, many ways.

For instance:

The anti-inflammatory aspect: When you shine a therapeutic grade red light on your skin, it stimulates a biochemical process that enhances your skin cells ability to create energy for themselves, the overall skin layer, and for other parts of the body. Because all our organs are an interconnected web, when the cells on the skin heal and rejuvenate themselves better and faster thereby bringing about a healing effect, this effect passes on to other areas and parts of the body.

One aspect of the healing effect red light therapy has on the cells of the skin is that it significantly reduces inflammation and stress. Again, when you shine a therapeutic grade red light on skin cells, it enhances their functionality. What this means is that the cells receive better energy, blood and oxygen flow, which means an enhanced ability to heal, repair, and rejuvenate.

To be specific, a research study conducted by Dr. Michael Hamblin noted, *"skin cells consider the wavelengths used for red light and near infrared light therapy a form of mild stress. When you expose these skin cells to this form of mild stress, it activates their need to self-protect (and to protect you too). When the cells are in self-protect mode, they function more efficiently and are very adept at creating the antioxidants they need to fight any inflammatory effect caused by the mild stress."*

When it comes to reducing inflammation, red light therapy is so effective that one study published in Medical Science noted that RLT treatment have a very therapeutic effect on post-surgery inflammation. Reduced post-surgery inflammation leads to better pain relief, less wound swelling and irritation, and because of enhanced cell functionality, quicker healing and recovery times, which means with consultation, red light therapy after surgery can be a very beneficial form of therapy for wound management, one that also stimulates the cells into working better and healing quicker.

An oral research study conducted by Lee JH, et al concluded that using red light therapy on the mouth—exposing the periodontal cells to therapeutic red light—leads to decreased inflammation and an enhanced cells ability to stave off toxins such as P. gingivalis and E. coli, both of which are very common mouth toxins.

We cannot talk about inflammation—and the effect red light therapy has on its reduction, and leave out mentioning the effect this therapy has on musculoskeletal soreness or muscle soreness.

A research study published in 2008 noted that controlled heliotherapy—as a reminder, RLT is a form of heliotherapy, the controlled use of light for healing purposes—significantly reduces symptoms of muscle soreness. In another research study from Brazil, researchers concluded that using a red light therapy machine before workout led to a significant decrease of post-workout pain and muscle soreness/inflammation.

RLT Enhances Sleep

Better sleep means better rest and therefore more energy and increased productivity in your day. Red light therapy has proven effective and capable of improving your sleep patterns, which it does by normalizing and then enhancing your circadian rhythms and cell's ability to produce melatonin, the hormone whose primary production of is to help the body control your sleep-and-wake up cycle.

Morita T., Tokura H conducted a research study to determine the effect Red light therapy sessions during the day may have on the body's ability to produce melatonin. She determined that exposure to pure red light during the day led to enhanced melatonin production and a rationalized circadian rhythm. This congruency can only mean one thing: red light therapy sessions during the day will lead to better sleep and rest.

To enhance your understanding of how red light therapy enhances sleep, here is how the process works.

How this works

Let's start by agreeing on one thing: poor sleep patterns have negative short and long-term consequences and effects.

For example:

Poor sleep patterns affect hormonal balance, which then affects our mood/wellbeing. Poor sleep patterns also lead to low energy, which means decreased output and productivity, which by itself brings a ton of negative consequences such as an inability to achieve your aims.

Additionally, a sense of 'chronic tiredness' can lead to mental health complications such as anxiety, stress and depression, and their attendant conditions such as rapid weight gain or loss, increased risk of type 2 diabetes, and insulin imbalances that disrupt your body's ability to determine when it needs sustenance.

To compound the issue, some research studies have concluded that poor sleep can lead to decreased cognitive function such as a low attention span and decreased mental clarity.

Outlining how abnormal sleep patterns decrease the overall percentage of your wellbeing was important because it created a background that will help you understand how red light therapy ensures better sleep.

The first thing you should note is that our sleep-cycles are light dependent. This means that when it comes to sleeping, i.e. when it comes to how your body decides when it needs rest in the form of sleep, light is the determinant factor. The body uses natural (and manmade) light to determine when to produce the hormones it needs to ensure that the body [as a machine] has an operationally effective circadian system/rhythm.

Because our modern lives are full of artificial lights—think bulbs, screens, and such—this state can lead to a confused circadian pattern, one that throws your sleep cycle off balance, thereby making it harder for you to fall asleep naturally, rest well, and wake up feeling refreshed and rejuvenated.

Secondly, it's important to point out the negative effect some lights have on our circadian rhythm and ability to sleep and rest; for specificity, we shall concentrate on blue light.

Exposure to blue light is very common in our daily lives in the form of screens such as televisions, monitors, laptops, tablets, phones, etc. Since most of us have grown accustomed to blue light exposure in the evening and long into the night, our bodies have learned to interpret this to mean it should remain awake because it's light outside.

Various clinical research studies have demonstrated and concluded that having red light therapy sessions in the evening has the potential to improve your sleep patterns. The science behind this is as logical as logic is! Because pure red light has a lower temperature, exposing your body to such wavelengths of light in the evening tells your body that the light has changed to a low temperature one [evening/night] and that it should prepare for rest by producing the necessary hormone, melatonin.

In 2013, Taiwanese researchers sought to determine the state of brain activity (electroencephalography EEG) before, during, and after red light therapy sessions. They did this by analyzing the EEGs of study participants. From this research study, they concluded that exposure to a therapeutic, pure red light in the evening was an effective way to combat common sleep disorders.

In another research study published in 2014, researchers noted that after red light therapy, patients suffering from traumatic brain injury (TBI) experienced less PTSD, reported better sleep patterns and improved brain functionality.

In yet another Brazilian research study published in 2018, researchers concluded that red light therapy proved highly effective against sleep disorders and other no-sleep related complications such as migraines.

As mentioned earlier, part of the reason for its effectiveness as a form of "alternative sleep therapy" is the effect it has on the signals that trigger the production of melatonin. Because pure grade red and near infrared lights have a low temperature, exposure to them at ideal times of the day can have a pronounced positive effect on the production of natural melatonin.

The logical conclusion we can draw from the collective message from the various research studies we have talked about here is that red light therapy is an effective, alternative form of sleep therapy so much so that lying in pure red light in the evening can make it easier to fall asleep and sleep better too.

In specific relation to how red light therapy leads to improved sleep, research by Margaret N., Michael H. has shown that after 18, 1-hour, daily red light therapy sessions, patients suffering from traumatic brain injury slept for an hour longer and woke up feeling more rested and refreshed.

RLT has Weight Loss Benefits

First off, when it comes to weight loss, nothing beats good-ol' healthy eating and consistent engagement in some form of physical activity such as exercise. With that noted, we can do various ways to enhance weight and fat loss. Some examples include practicing intermittent fasting—eating within a specific hour window—taking weight loss supplements such as garcinia cambogia, and detoxing. RLT can also be effective as an alternative weight loss strategy you can couple with healthy eating and adequate exercise and use to enhance your body's ability to burn fat.

Various research studies have shown the positive influence red light therapy has on weight loss, gain, or retainment. In particular, researchers have noted that red light therapy is an effective, alternative weight loss strategy because its effects act upon the adipocytes cells, the cells in your body responsible for storing fat.

In 2011, Paolillo FR, Borghi-Silva A, et al. published a study detailing what the researchers discovered after subjecting study participants—women aged 25-55—to treadmill exercises and red light therapy. The study revolved around categorizing the participants into two categories: those who engaged in treadmill running, and those that did treadmill running and two sessions of red light therapy each week, with the study-measure being to determine how red light therapy influences cellulite.

These researchers demonstrated dramatic differences between the two study groups. The starkest of these differences was that the group that engaged in treadmill/cardio exercises and two red light therapy each week showed dramatic improvement of thigh cellulite; the researchers then concluded that engaging in exercise and RLT leads to an aesthetically pleasing body and increased fat loss.

In 2012, the International Journal of Endocrinology—the study of medicine in relation to hormones and glands—published a research study where researchers concluded that light influences our eating patterns and that this can itself influence our hunger and satiety levels.

These researchers concluded that when your hunger and satiety hormones and signals are out of whack', the effect is a decreased ability to regulate when [and when not] you eat. This decreased ability can lead to the development of condition such as binge eating, and various hormone-regulations conditions such as type 2 diabetes and a cellular attachment to storing fat in the form of lipids. Red light and near infrared light therapy increases cellular activity, which effectively means that as cells work better, their ability to release fat lipids increases. When this happens, i.e. when the cells release fat lipids, the effect is weight/fat loss. Coupled with healthy eating and exercise, red light therapy can dramatically improve your weight loss journey.

To enhance your understanding of how red light therapy enhances weight loss, here is how the process works.

How this works

Let's recap what you now understand about red light therapy. You know that this alternative therapy works by delivering pure red light on the skin and that this causes increased cellular activity. You also know that once red light stimulates increased production of ATP, it ensures your cells operate optimally and as effectively as possible.

Red light therapy works so effectively as an alternative form of weight loss therapy because it activates enhanced activity in adipocytes, the body's fat regulating cells. When red and near infrared lights contact the skin, the resultant biochemical reaction triggers increased activity in these fat cells that when activated, cause the cells to release any excess fat (lipids) they may be carrying unnecessarily.

In addition to this, researchers have concluded that the effectiveness of RLT as an alternative weight-loss management system comes from its ability to regulate the production of various hunger hormones [but specifically ghrelin and leptin].

Tons of research findings fully support red light therapy as an alternative weight loss strategy/system that coupled with healthy eating and adequate exercising, can have amazing fat loss and increased metabolism benefits.

One part of wanting to lose weight is sculpting the 'body of your dreams.' To this end, you may use various therapies at your disposal such as various body sculpting techniques and in some cases, targeted surgeries.

There is scientific evidence to prove that using RLT with weight loss as your intent and coupling this therapy with other good-health strategies can greatly improve your chances of losing even the stubbornness of fat and help you contour your body to get the body you have always wanted.

Handheld RLT devices are an especially effective way to 'contour' the body simply because they allow for better targeting and maneuverability. By allowing for better targeting of specific 'fatty' areas such as love handles, thighs, and underarms, red light therapy allows for greater and targeted cellular stimulation of the aforementioned fat cells. By using pure red light to influence increased activity in areas that tend to carry fat, your body releases the lipids within cells, which then makes the weight loss process easier.

In one study conducted in 2011 and results published in the Journal of Obesity Surgery, after a double-blind study, researchers concluded that red light therapy could lead to a slimmer waistline. The study involved allowing participants access to red light therapy sessions of 635-80nm for a four week period. The researchers focused on the effect these therapy sessions had on the girth of the waistline. Their study concluded that exposure to red light at the aforementioned wavelengths has great potential as a waistline sculpting tool.

Various other studies support the findings of the aforementioned study. One such study appears in <u>Lasers in Surgery Medicine</u>. In this research study, researchers sought out to know the effectiveness of RLT as a targeted weight loss tool—how effective it is at helping one lose weight at targeted areas. After offering participants red light and near infrared light therapy sessions, the researchers found that these led to a significant loss of body fat in key areas such as the thighs, underarms, etc.

In another study published in 2013, researchers determined that exposure to pure red light at 635nm led to contoured thighs and hips. Some participants in this study saw a reduction of up to 2.99 inches of their augmented body size.

In <u>a recent study [2018]</u>, researchers sought to determine the specific effect red light therapy has on loss of body fat especially during endurance training. After the study, researchers concluded that compared to the placebo, participants who had red light therapy sessions pre-exercise experienced greater body fat synthesis and loss—their bodies burnt fat much faster during endurance training sessions and long after.

Red light and near infrared light therapy has also proven very effective as an obesity management tool. After assessing the results of their research findings from offering 64 obese women aged 20-40 red light therapy, Brazilian researchers concluded that coupling this therapy with exercise has the potential to lead to greater weight loss. The researchers noted that, *"when obese women engage in light therapy and exercise, in addition to increased weight loss, they displayed significant metabolic flexibility."*

If you go digging for it, there is no doubt that you will find plenty of research findings that support the use of red light therapy for weight/fat loss. One of the great things about using red light therapy as an alternative, weight loss therapy is that it gives you more control and targeting, and it's safe to do at home.

Today, the marketplace has a wide offering of effective, handheld, or at-home red and near infrared lights devices you can use when you wish—keeping in mind the guidelines your physician recommends—and more specifically, ones you can use to target specific problem areas such as those love handles.

Because pure red light is safe, and because there are plenty of research findings to support its effectiveness as a weight management tool, provided you consult your physician, there is no reason why you cannot use red light therapy as a weight loss and body-sculpting tool.

RLT Enhances Muscle Recovery and Performance

From everything we have discussed about red light therapy thus far, it's relatively easy to see why red light therapy could have positive effects on performance and muscle recovery.

First, exposing various layers of the skin to pure red light increases cellular metabolism. What does this translate to in terms of performance and muscle recovery?

Think of it this way.

By undertaking red light therapy sessions and activating the cells into producing more ATP/energy, activity in the cells increase, and because the cells have access to cleaner, efficient energy, their performance increases and they function more effectively.

Moreover, because stimulating cells in this manner leads to better circulation and cellular communication, engaging in red light therapy can have positive effects on cellular and overall body and brain performance.

Here is how this works:

RLT and performance: How this works

There's a reason why top athletes and professionals are advocating for red light therapy as a way to increase physical performance, grow muscles and speed up recovery after training: because the therapy is that effective. Because tons of research studies have shown the effectiveness of RLT as a performance and recovery tool, Olympians, NFL all-pros, NBA teams, boxers, and professional gyms have taken to using it.

In a 2011 study where researchers set out to determine the effect red light and near infrared light therapy has on strength training in healthy men, they noted that coupling strength training and light therapy leads to improved muscular performance and recovery especially when compared to the results displayed by the test group.

In a 2014, controlled study, researchers noted that red light therapy leads to increased grip strength. In this specific study, researchers noted a 52% increase in grip/hand-based exercises during strength training sessions.

In a scientific trial conducted in 2016, researchers sought to determine how red light therapy sessions would affect strength training in men aged 18-35. The researchers concluded by saying that men who engaged in strength training and red light therapy sessions had increased mobility/torque and strength when engaging in leg-based exercises such as leg presses and extensions. They noted that engaging in phototherapy before strength training sessions leads to enhanced strength and that it is highly beneficial for post-workout recovery.

Numerous research studies have also proven RLT effective at enhancing endurance. Improved endurance means you can work out for longer irrespective of the type of exercise you undertake.

As an example, in 2018, a controlled trial sought to know how red light therapy influences endurance in healthy men and women using treadmills as their preferred exercise machine of choice.

The researcher concluded by noting that, *"engaging in light therapy before exercise increases oxygen uptake, increase how long volunteers could exercise before hitting the point of exhaustion, and significantly improves fat loss."* These researchers then proposed that engaging in red light therapy before endurance training can significantly increase endurance by up to three times.

In 2018, Brazilian researchers and futsal players collaborated in a triple blind study. First, Futsal is a type of indoor soccer popular in Brazil. Unlike soccer, because the playing field is smaller, the game is more challenging and demands better endurance.

After allowing pro-Futsal players access to red light therapy sessions before matches, and analyzing their findings, the researchers determined that Red light therapy participants could out-endure the placebo group. Their specific conclusion was that: *"Red light therapy significantly improved how long a player could stay on the pitch, and had an overall positive effect on the various bio-chemical markers segregated for monitoring pre-exercise."*

Two other trial studies conducted by the same team have backed the results of the aforementioned Futsal study. In <u>one study</u>, researchers sought to determine the effect red light therapy would have on the performance of professional cyclist. In this study, researchers concluded by noting that red light therapy sessions before professional cycling bouts increased how long it took for a pro-cyclist to get tired to the point of exhaustion.

In <u>the second research study</u>, researchers set out to determine the VO2 kinetics during cycling test—this term means the rate at which the body responds to increased oxygen demand during exercise. This study also displayed similar results: that RLT pre-cycling tests increases time to exhaustion and oxygen uptake.

Various research studies have concluded that the other way by which red light therapy improves performance and muscle recovery is by enhancing speed—this is specific to running.

<u>In 2016</u>, researchers decided to study the key markers that assess physical performance in pro rugby players. After concluding their research study, they presented findings that illustrated the effectiveness of Red light therapy. They specifically showed that RLT leads to faster running, improved sprint times, and better muscle recovery in rugby players.

<u>In 2018</u>, Brazilian researchers conducted another study that sought to know if red light therapy sessions before running

workouts would lead to efficient, faster running. The double-blind study had three groups: a placebo and control group that did not receive red light therapy session, and men who received these treatments.

From their results, the researchers noted that the men who engaged in red light therapy were the highest performing: they ran faster and for longer, and had higher peal velocities compared to men in the other two groups.

Before we move on, we should mention that various research studies have shown that red light therapy works effectively as a muscle growth/enhancement tool.

One such 2010 study conducted by Kelencz CA, Muñoz IS, etc. and later published in *Photomedicine and Laser Surgery*, researcher sought to use both genders to determine whether red light therapy leads to less fatigue and more muscles. They concluded that the group that engaged in red light therapy displayed significant increases in muscular activity and strength.

Another research study published in The American Journal of Physical Medicine and Rehabilitation noted that RLT and near infrared light therapies can lead to muscles size and bulk; it does this by promoting cellular growth of healthy muscle tissues and muscle hypertrophy.

A unique study published in the European Journal of Applied Physiology used two groups, a control/placebo group, and another group that engaged in red light therapy, to determine the muscle and strength growth and differences between the two groups. They concluded by noting that the muscles of the group that coupled exercise with red light therapy were thicker and stronger (by up to 50%).

Given what we know about therapy with red light and near infrared light therapy, it is easy to see why red light therapy would be effective for muscular recovery, strength, and performance. When cellular activity increases, which is what happens when you expose the human body/skin to pure red or near infrared light at the right wavelengths, inflammation decreases while all other cellular functionalities increase.

Moreover, because of its anti-oxidation effect on the cells, muscle fatigue reduces, which when it does, leads to improved performance and an increased production of heat hormones, which are highly important proteins that ensure cell health and operation by safeguarding cells against chronic oxidation, mutations, decay, and eventual death.

Light therapy also leads to improved blood flow. When your muscles receive more blood, they perform better, endure more, and heal/rejuvenate faster.

RLT Improves Inflammation and Joint Pain

The body uses natural and pure red light to increase activity in the lymphatic system, the body's waste and toxins management system. Scientific research has shown that increased activity within this system reduces inflammation and swelling.

What does this mean?

It means because of its effectiveness against pain and inflammation, using red light therapy for inflammation-based ailments such as arthritis can be beneficial, and as various research studies have shown, highly effective because it acts on the cell nuclei thus treating the root cause of the problem.

In one photomedicine research study conducted in 2018 by Brazilian researchers, results of the study indicated a significant decrease in cytokine after red light therapy, and concluded that red light therapy can accelerate inflammatory response by stimulating cellular regeneration—remember that RLT activates your cells ability to create ATP, the most important molecule.

Numerous studies have proven RLT safe and effective as a joint pain management system and as a way to manage arthritis. For instance:

RLT effects on Osteoarthritis: Numerous studies have sought to determine whether photomedicine is an effective way to manage knee pain and Osteoarthritis.

Given that using a red light device is an effective way to activate cellular metabolism, it makes sense that red light therapy would be an effective way to manage knee related pains. In two separate Brazilian studies conducted in 2018, researchers concluded that coupling light exercising and stretching with red light therapy was more effective at treating the ailment (osteoarthritis).

One of the research teams specifically noted that, *"after red light therapy and stretching or exercise, study participants illustrated significantly reduced levels of knee pain and increased mobility after 90 days of light stretching."*

Researchers have also proven RLT effective as a general pain management tool. After reviewing 11 clinical trials on red light therapy and photomedicine, the review team at the Australian Journal of Physiotherapy concluded that across the clinical trials, overall, participants displayed significant reduction in joint pains including general knee pain.

RLT has also proven very effective at managing varied forms of joint pains including wrist and hand pains. A review study published by Paolillo AR, Paolillo FR, et al in Lasers in Medical Science concluded that light therapy had proven effective at treating hand pain (hand osteoarthritis) in women, most of whom reported significant reductions in pain and increased mobility.

In 2016, Baltzer AW, Ostapczuk MS, Stosch D. sought out to know whether red light therapy—low level laser therapy—has any positive effects of Bouchard's and Heberden's osteoarthritis, a bony outgrowth and a swelling condition respectively. The study had 34 participants. The research team noted that red light therapy led to significant improvements in hand mobility, reduced swelling, and very positive effects on ring size.

Scores of other studies support red light therapy as an effective way to manage spine pains caused by various conditions such as ankylosing spondylitis. In one such study published in 2016, researchers noted that participants who combined stretching with red light therapy displayed improved mobility and significant reductions in spinal pain.

From the various research studies we have discussed here, we can conclude that plenty of research findings have proven red light therapy effective and beneficial in many ways.

Because it's non-invasive and drug free, red light therapy is normally safe to use for most people—of course they're contraindications. Moreover, because of the controlled nature of the therapy, more so in terms of the wavelength of the light used during therapy, red light therapy has thus far shown no long-term negative side effects.

In addition to the above, we can also infer that the popularity or red light therapy has seen significant growth as the technologies driving it advance. Today, thanks to handheld RLT devices and treatment centers around the world, anyone can use this alternative therapy.

Because red light therapy treatment centers have varying procedures and regulations, we shall deconcentrate on treatment within the confines of these centers and instead concentrate on using Red light therapy at home.

Red Light Therapy at Home

Because photobiomodulation (or light therapy) has advanced significantly, we have many red light therapy options at our disposal. As noted, you can receive this kind of treatment at professional spas or alternative therapies clinics.

In some instance, you can even seek professional heliotherapy treatments from your local beauty salon that may have a modified tanning bed or professional red light device that allows for full body treatments—most of these devices are too expensive for ordinary home users. In most of these establishments, a session of red light therapy costs $25-$100 per session.

Earlier, we mentioned that because of advancements in technology—especially advancements in photobiomodulation—red light therapy is now safe to use at home and in fact, at-home red light devices are infinitely more popular compared to in-clinic treatments or therapy sessions.

Once you consult your physician and make the choice to start using red light therapy at home, the main challenge you will encounter is that of choosing a device. This is because the devices available for at-home red light treatments are plentiful.

To choose a red light device that suits you, you need to keep various things in mind. This chapter looks at everything you need to know as you go about making the important decision of choosing a red light therapy device for at-home red light therapy sessions.

Choosing an RLT Device: Important Considerations

Now that you have a firm understanding of how red light therapy works as well as its many benefits, the next logical step is to start using this therapy so that you can distill the many benefits we discussed earlier. To do so, you will need to purchase an RLT device.

As you think of which RLT device suits your specific needs and aims, keep the following important considerations in mind:

#1: Consider wavelength and intensity

We have talked a lot about the wavelengths suited for red light therapy. You cannot overlook the wavelength of an RLT device you are considering purchasing and using simply because using the right wavelength of red light is paramount.

Most at-home red light therapy devices have a wavelength of 630-700nm. That noted, if you are looking for it, perhaps because a health professional or dermatologist has advised its use, you can also find devices with a wavelength of 600nm—the light from such a device is more orange than red—and even devices of higher wavelength of 800-900nm—these are what we would refer to as near infrared.

When it comes to choosing a device of the right wavelength, the best thing is to go for devices in the mid-600-low/mid800nm range; devices within this wavelength are the ones that have the most impact on the ATP generation process we discussed earlier—most of the benefits we discussed earlier were from use of devices within this wavelength.

*Tip: As you research which device to get, concentrate more on devices that offer the above wavelengths (660nm-850nm) because devices of these wavelengths have the most impact on the cells.

In RLT, intensity defines how much energy a device delivers to the cells. Here, the best thing to do is to favor devices that have the ability to deliver 4 to 6 Joules/cm2—but up to 40 Joules/cm2—to the cells of the body. Devices that offer these kinds of intensity have proven the most effective in many research studies, including ones supported by NASA. Devices of higher intensity are especially effective for conditions such as joint pains and inflammation.

Here, avoid devices that do not explicitly display their intensity output; most legitimate devices will display the irradiance—energy output—of their devices.

*Tip: When considering the intensity of a specific device you are considering purchasing, look for the irradiance, which most manufacturers will represent as energy/Joules/cm2 per minute, or as mW/cm2.

If a device has displayed its irradiance, to determine how much energy intensity it delivers to the skin, use the following formula: *I (for Irradiance) X 60 (time seconds) ÷ 1000*

For instance, if a device indicates it has an energy output of 15 mW/cm2, we would calculate its energy intensity as 15 mW/cm2 X 60/1000=0.9 Joules/cm2 per minute.

In the above example, you should note that such a device would not be very effective because as we have mentioned several times in this section, the best devices, the ones that offer the benefits we mentioned earlier, are ones that deliver 4-6 Joules/cm2 per minute. Some devices have higher intensities—some deliver-to-the-skin 60 Joules/cm2.

The intensity of a device is an important consideration because it determines the duration of therapy. A lowly powered device means therapy sessions will be longer while a high-powered one will translate into deep tissue stimulation in less time.

#2: Consider your lifestyle

After considering the wavelength and intensity of a device, the next thing you have to consider is how red light therapy fits into your present lifestyle as well as whether the device you are favoring complements this lifestyle.

Here, consider how much time you have to dedicate to at-home therapy. Ideally, you should opt for a device that delivers optimal wavelengths and intensity of red light fast and conveniently. This means if you don't have too much time to dedicate to RLT, a small device may not be ideal for you.

The other thing to consider here is the convenience of using the device. When thinking about this element, your first consideration should be the areas of the body you intend to treat. For instance, if you want to treat the face and just the face, you can opt for an RLT therapy mask. On the other hand, if your intention is to treat the entire body, you are better off going with a bigger device such as a specially created tanning bed.

#3: The Treatment/coverage area

You cannot fail to consider the coverage area; as mentioned, a smaller device will treat a smaller area.

Here, how much area of the skin a specific RLT device can cover or treat will depend on three primary factors: its dimension, the angle of the refracting lens, and the distance between the light and the skin surface.

On dimensions, the RLT device market has plenty of RLT LED panel offerings to suit your specific needs; some are small, some large, others broad, and others narrow. Here, your budget will largely determine what kind of a device you get. Devices that have larger LED panels cover a greater treatment area, have more energy output, and are thus likely to be costlier. Your budget allowing, you should always opt for a device that has a bigger LED panel as this means more power density and coverage.

While still on dimension, factor in the areas you intend to treat because for instance, a device that covers the face will be ideal for use on the face only, which means if you opt for such a device while at the same time intending to use it on other areas, you will run into some challenges.

On the angle of the refracting lens, this is the lens placed over the overall LED panel to ensure final dispersion of the red light is even. This lens is very important as it ensures the effectiveness of the light emitted. Here, most manufacturers use secondary lenses or bare LEDs.

Secondary lenses have a refracting angle of 30, 60, and 90 degrees. The effect of this is that the angle of refraction affects the coverage area. A secondary lens with a smaller refracting angle means higher concentration in one area but diminished coverage in other areas. Most good devices use secondary focusing lenses of a higher angle (60 degrees and maybe more); this ensures effective coverage and output.

On distance from the skin surface, the main thing to keep in mind is that the further away the light is from the surface, the wider the surface area it will cover. In some instance, especially in instances where the light in question is low quality, the LED panel being further away from the skin surface means lowered intensity; be mindful of this.

#4: **Manufacturer and warranty**

When comparing RLT devices, you cannot overlook comparing various manufacturers, their level of experience, and the kind of warranty/guarantee offered.

Here, to get the best bang for your buck, concentrate on manufacturers who have a proven record of accomplishment of innovation in this field and who offer a warranty they are willing to stand by. For instance, companies such as Joov and PlatimumLED are leading innovators in this field and their devices are some of best in the world.

The strength of a brand/manufacturer dictates is consumer trust. As you think of which RLT device manufacturer to patronize, think about the warranty period and make it a point to read reviews especially those related to how the manufacturer handles the warranty claim process.

#5: **Consider your aim**

As mentioned severally, how you intend to use your RLT device is an important consideration, one that you would do well not to overlook. Here, if your aim is to ensure the light seeps deeper where it can affect deep tissue, near infrared light devices are your best bet. On the other hand, if your intention is to use red light therapy for skin enhancement purposes—as opposed to treating other conditions such as joint pains or arthritis—red light devices will be more to your liking.

For general red light therapy, opt for a large LED panel—this way, you can have better coverage—that can deliver 660nm-850nm. Devices that deliver a mix of these wavelengths can work for a number of uses such as joint pains, as a therapy for skin conditions, and even for fat loss.

To use red light for enhanced skin and hair loss reversal, lights of 660nm work best. To target joints, muscles and tendons, and even organs, near infrared or pure light devices of 850nm will be ideal.

If your intent is to concentrate on using red light therapy for brain enhancement or as a remedy for anxieties, depression, and other psychosomatic conditions, near infrared light devices are your best bet.

If you keep these considerations in mind as you research RLT devices, you should be able to narrow your list of devices to a few that match your exact red light therapy use.

Having looked at how to choose a red light therapy device, let us talk a bit more about dosage.

Photobiomodulation/Red Light Therapy Dosage

To use red light therapy effectively, you need to pay attention to the dosage; in most cases, this means paying attention to the energy intensity that your RLT device can delivery at various distances away from the skin surface as well the duration of therapy sessions.

Red light therapy dosage depends on the intensity of the light. The higher the density/intensity, the more concentrated the red light effect at given distances. For instance, some red light and near infrared light devices will have a power density of 450-600 mw/cm^2 at a distance of 5cms from the point of contact/skin.

On dosage using power/light density, the fundamental thing to note is that clinically, light devices that have a density of over 200mw/cm2 are not ideal for treating skin conditions. However, when used in a controlled manner and for shorter periods, such devices can be ideal for deep tissue stimulation—think treating joint and back pain, arthritis, etc.

After determining your device's power density, you can use this to determine how far away to place the device, we call this the light therapy range. The idea here is to ensure you place your red light device or LED panel at a distance that allows for appropriate power density. With devices that have a 10-200mW/cm2 range, this distance is usually 0-35cm.

The formula

To calculate the ideal dosage for photobiomodulation, use the following formula: *Dose = Power Density x Time x 0.001*

As mentioned earlier, the power density is the most important consideration especially for at-home red light therapy use. Without knowing a device's power density, calculating the ideal dosage for optimal effect will prove difficult.

Again, most high quality red light and near infrared devices will clearly display their light intensity/density—some manufacturers may choose to indicate the photons within a specific area of space.

To cover a bigger area, some RLT devices such angled output LED devices capitalize on how light spreads. In such devices, the further the light spreads out, the more its coverage area; the compromise to this is that the further away from the base source, the weaker the intensity/density.

The higher the power density or light intensity a device has, the less time you will need to spend under it; most of the benefits we discussed earlier are from clinical trials and studies using lights of a power density of 10mW/cm2-200mW/cm2.

Assuming your intention is to determine the dosage for an area of the skin measuring 40cm X 40cm or 1,600cm2, and that the red light device you intend to use has 200 watts or

200,000Mw, the device would have a power density of 125mW/cm2, which is a great power density.

Once you have the power density at various distances as well as your preferred dosage in J/cm2, you can use the following formula to calculate the duration of the therapy in seconds: *Time = Dose ÷ (Power density x 0.001).*

Red light therapy dosage normally depends on the intensity/power density of the device in question, the duration of dosage (how long you stay under the light), as well as the distance from the surface. The unit of measure used to display the dosage is Joules per centimeter squared (J/cm2). Again, the greater the power density, the lower the dosage or rather, the higher the power density a device has in relation to distance from skin surface, the lower the application time (in seconds).

Conventionally, most of the red light therapy benefits we discussed earlier manifest when the distance from surface—the distance between the red light device and the surface of the skin—is 20mW/cm2 on the lower end and 200mW/cm2 on the higher end for larger devices. Placing the red light therapy device further away from your skin will call on you to compensate by increasing the duration of the treatment.

The ideal dose to aim for

An ideal dosage is one of 3J/cm2-50J/cm2. However, the ideal dosage will depend on your aim for using red light therapy.

One thing to keep in mind is that application time, i.e. how long you apply the red light to your skin at a specific power density and distance, will influence the dosage.

Additionally, dosage will depend on the condition you intend to treat. For instance, to treat acne, an effective dosage is 5-96J/cms2. For back pain, a dosage of 40-100J/cm2 is ideal. The following resource page has ideal dosages for common conditions treatable using red light therapy: Dosage resource page.

For dosage, the guidelines to keep in mind is to have a high powered red light therapy device so that when you move further away from it, you can still ensure that larger parts of the skin get the right intensity of light for cellular activation.

Additionally, the number of treatments per week will largely depend on the condition treated. With that noted however, the general guideline is 2-14 treatments per week— depending on the condition and the dosage—with mornings and evenings being great times for red light therapy.

When starting out with red light therapy, start with low dosages at healthy distances and then gradually increase the dosage as the surface of the skin accustoms to red light usage. Moreover, be mindful when treating highly sensitive areas such as the face; with these areas, be conservative with the dosage.

To recap, to calculate the right dosage so that you can get the best out or red light therapy, measure your device's power density (mW/cm2) at different distances—use a solar meter to measure this. Once you have the power density, calculate the dosage by multiplying it by time in seconds; for general use, go for a dosage of 3J/cm2-50J/cm2.

To avoid over dosage, which may lead to negative effects because with red light therapy, more is not usually better, its best to stick to the recommended dosage range.

Now that you have a firmer understanding of red light therapy dosages, the next thing we are going to talk about is actual use of the device in everyday life and the various contraindications to be aware of or to keep in mind.

Red light Therapy Tips

Having used the information contained within the last two sections to choose an ideal red light therapy device and the right dosage depending on the condition you want to treat, the next step is to continue using red light therapy consistently as advised by your dermatologists or physician.

Here, as you go about using your device—or seeking red light therapy treatments at a treatment center—keep the following essential tips and strategies in mind so that you can get the best out of your therapy sessions:

#: You can use RLT more than once daily

You can have more than one red light therapy session a day; the only thing to keep in mind is the optimal dosage and the upper limit of your dosage. For clarity, the ideal dosage is 3J/cm2-50J/cm2 or higher for deeper tissue stimulation.

As mentioned earlier, in red light therapy, the concept of more being better does not apply. In fact, overdoing therapy sessions—maybe staying under the light for far too long or having far too many sessions in a day—can have some adverse effects on the skin, not as adverse as overdosing on UV light, but undesired nonetheless.

With red light therapy, the keyword is balance: *treatment should be a balance of using a device that has an adequate power density, placed at an ideal distance to surface/skin, and used for the right amount of time daily or weekly as advised.*

Just as there is overexposure to red light therapy, there is underexposure too. Too many red light therapy sessions negate the positive effect of red light therapy; on the other hand, underexposure means no benefits at all.

For the best effect and the most benefits, aim for balance, an optimal dosage that allows you to derive the many benefits of red light therapy. Many of the available, high quality, red light and near infrared lights device come with instructions on the power density at various distances and the ideal dosages—amount of time to stay under the device for a treatment—for various conditions.

Keep in mind that your ideal dosage for a day will vary depending on your goal, the power density of the device in use, and the time spent under the radiance of the pure red light. As long as you are not overdoing it, you can use your RLT as many times as necessary in a day.

#: About protective gear, makeup, and clothing

Today, the most common red light therapy devices are red light facemasks or red and near infrared lights that you can shine on various parts of your skin or place further away from you depending on the power density. Inadvertently, this brings up the question of protecting sensitive areas of the skin such as the eyes perhaps by wearing eye-protecting gear/goggles.

On protective eye gear, many experts and researchers in this field have concluded that red light therapy at a wavelength of mid-600nm can be beneficial to the eyes. This means that if you are using a device calibrated to the aforementioned wavelength of red light, you do not have to wear eye-protecting gear.

However, if you have specific light sensitivities, it is best to acclimate your eyes to the red light by using it in a well-lit room or looking at and away from the device in short successions. You can also shut your eyes for a few seconds, open them, and look at the red light for a few seconds before closing your eyes again and repeating the process a few more times or until your eyes accustom to the red light.

Moreover, if your device offers more than the aforementioned wavelengths (mid-600nm-700nm)—perhaps you have a red light or near infrared device of 850nm and above—shut your eyes or wear eye-protecting gear during therapy sessions so that you can protect the eyeballs from overheating during treatment. Most red light or infrared red facemasks have internally built protective eye mechanisms that allow you to keep your eyes open throughout the therapy session.

On other forms of protective gear such as clothing, the important thing to remember about red light therapy is that the more skin you expose to red light at the right density/intensity and for the right amount of time, the more beneficial the therapy.

If you have a large enough red light or near infrared device such as a full body device, you do not need protective gear and for therapy sessions, you can in fact strip down to your undergarments. Depending on the density of the device at various distances, wearing light clothing may be necessary.

On makeup and other applicable beauty products such as lipstick and the likes, you can still benefit from red light therapy with your 'makeup' on, but for the best result, its best to clean the skin before a therapy session. Most high quality red and near infrared light devices will come with a set of usage instructions. In specific, most facemasks will have a detailed procedure indicating how to treat the skin before and after a red light therapy session.

On using your device after a bath, most of the available devices use high-grade electronics parts; they are still electronics and if your intent is to use a red light device after your morning or evening bath, its best to dry off first.

#: **On treatment guidelines**

As mentioned earlier, the correct dosage will depend on your device and the intended area of treatment/condition. With that noted, most available red light and near infrared light devices will come with a set of treatment instructions or guidelines depending on the device's light density from various distances.

Most high quality, at-home red light devices will advocate for sitting or standing 4-6 inches away from the device; this is because 4-6 inches away from the device—assuming the power density is adequate—is the clinically proven ideal distance from surface.

Most available at-home red light therapy device and even treatment centers will also have some form of guideline on how to get started. In most instances, both at-home devices and treatment centers will advocate for easing yourself into the therapy with short sessions and then increasing the duration. For instance, some manufacturers will recommend easing yourself into the therapy and usage of the device with 1-2 minutes therapy sessions on a smaller/targeted area, and then building up to 10-15 minutes sessions within a specified window of time.

Most RLT treatment centers and devices advocate for 10-15 minutes sessions; this ensures optimal exposure. However, as we have mentioned several times, the correct dosage is a balance based on the power density at given distances.

When using a targeted device, one that allows you to target a specific area of the skin, once you use such a device on a specific area of the skin, say the thigh, it is best to avoid treating that area for at least 6 hours. With full body devices, you do not have to worry about this because these shine red light on large portions of the body and therefore only necessitate one or two therapy sessions per day. Keep in mind that more red light does not necessary translate to enhanced benefits, overexposure may in fact negate the intended benefits.

Most at-home red light therapy devices will advocate for daily use. This is congruent with what the research notes, which is that the benefits of red light therapy come about from consistent use of the therapy.

As mentioned, you can have more than one therapy session a day provided you keep in mind the upper limit of your dosage. The more consistently you use your device, the greater the impact on cellular activity and regeneration. Like working out, red light therapy treatments are more beneficial when you create and stick to a regular treatment schedule. Moreover, sticking to a consistent and regular schedule ensures you derive from the practice long-term benefits.

On the best time to use red light therapy, morning and evenings are ideal times because you can schedule red light therapy into your morning and evening routine—for energy rejuvenation in the morning and as a way to relax and unwind in the evening.

With the above mentioned, anytime is a good time for red light therapy. If you have free time in the afternoon, a red light therapy/treatment session then will just be as effective as one conducted in the morning or evening.

What time of day you use your red light therapy device will largely depend on the benefits you want to accrue. For instance, if the idea is to speed up muscle recovery after exercise, scheduling your red light treatment after a workout will have the most effect. On the other hand, if the idea is to improve sleep, scheduling your red light therapy session as part of your wind down evening/night routine makes more sense. On the question of when is the right time for a red light therapy session, the answer is do what feels right for you depending on the benefits you want to distill.

#: **Duration to benefits**

How long it takes for red light therapy treatments to manifest will vary based on factors such as consistency, proper dosage, and various other factors such as distance from surface as well as the area treated or targeted.

With that mentioned, it is possible for intended benefits to manifest after a few weeks of consistent use of red light therapy. In some instances such as using red light therapy for joint pain and inflammation, some people have reported experiencing the benefits of red light therapy after a single therapy session.

For conditions such as wound and muscle recovery, benefits can manifest clearly after 2-4 sessions. However, to treat deeper-rooted conditions such as hair loss, fat loss, joint pain, arthritis, and the likes, benefits may take a bit longer to manifest.

The fundamental element to note here is that as long as you are using the correct dosage—a red light device of correct intensity, placed away from the skin at the right distance, with adequate exposure in terms of time—the benefits sought eventually manifest irrespective of how long it takes.

Risk, Side Effects, and Contraindications

Red light therapy is generally risk free and various reputable institutions such as the American Academy of Dermatology have classified RLT as 'safe.' This is congruent with what we have learned about pure red light, which is that unlike harmful UV light, it has no negative effects on the skin. Because the therapy is also non-invasive and drug free, it does not cause any damage to the skin.

As we have stated severally, with red light therapy, more does not equals better. Although over exposure will have dismal negative effects on the skin, you should keep in mind that overdosing on RLT negates the effects of cellular activation and increased ATP production.

With all the above noted, red light therapy does have some risks in some instances such as when treatment is on a skin that is sensitive to light. For example, using Accutane to treat acne increases your skin's sensitivity to light and therefore, using red light therapy while still using Accutane is likely to lead to scarring of the skin on some individuals.

Additionally, if you are highly sensitive to sunlight or are using anything that makes you sensitive to sunrays, its best to consult a qualified professional before using red light therapy.

Side effects from red light therapy usage are dismal and rare; in fact, many clinical trials have not noted any major side effects. Nevertheless, minor side effects such as eyestrain, headaches, and irritability can occur short-term. Most of these side effects are not explicitly from the use of red light therapy; rather, they are an effect of the glare of the red light, which may at times be intense.

As long as you acclimatize your eyes to the red light as illustrated previously, i.e. closing your eyes or wearing protective eye gear, which is a personal choice, you should be able to bypass most of the side effects. Additionally, aim to avoid too many instances of staring directly into the red light source.

In some studies, use of light therapy on individuals diagnosed with drug resistant, non-seasonal depression has shown to aggravate the condition into hyperactive mania. When such is the case, it is best to consult a qualified professional and to seek treatment for the condition before embarking on red light therapy.

More importantly, to avoid side effects, seek professional medical advice before you start using red light therapy.

NOTE: Worth mentioning here is that compared to red light therapy, near infrared light therapy, i.e. treatment with red light on the higher side of the wavelength spectrum, is more likely to cause side effects. This is because near infrared devices also emit thermal energy, which can lead to issues such as skin overheating and burning/thermal burn.

With near infrared light devices and therapy, aim to exercise proper care and caution; additionally, be strict with dosage and the distance from surface depending on the power density of the device in question.

Contraindications

In the following instances, it is best to consult widely before using red light or near infrared light therapy:

1. During pregnancy—although red light therapy is safe— even the FDA has classified it as safe for long-term use— pregnant women should consult their physicians before using photobiomodulation.

2. Epileptic patients–If you are epileptic, consult your physician before embarking on light therapy treatments.

3. After cosmetic filler procedure or Botox, avoid using red light therapy—the skin is more sensitive and thus much likely to scare. However, because red light therapy is effective at healing wounds, its best to consult a physician.

4. If you have a history of skin related cancers, consult your physician before using red light therapy or other forms of photobiomodulation.

5. If you suffer from Systemic Lupus erythematosus, it is best to avoid red light therapy and to consult widely before embarking on its use.

The most important contraindication is that if you are using any medication that makes you sensitive to light—we call these photosensitizing medications; examples include various antibiotics, some antipsychotics, melatonin, etc.—as a matter of importance, avoid red light use or consult your physician before using red light therapy.

Conclusion

We have looked at everything there is to know about red light therapy; from what it is, how it works, the benefits of the therapy, how to go about red light therapy including how to choose the correct device and the correct dosage, as well as the tips to keep in mind in terms of treatment guidelines.

Now that you know all this, keep in mind that red light therapy and all forms of photobiomodulation or heliotherapies are still alternative. This means unless advised so by a qualified professional, you should avoid any situation that calls on you to use red light therapy as the only way to manage a deeper condition. As we mentioned severally throughout this book, red light therapy seems to work well when coupled with other wellbeing strategies.

If after reading this book, you do decide that the benefits are well worth it check with your dermatologist or physician, and then seek out a therapy session or device that allows you to expose the skin to the right density of red light for the right amount of time depending on the condition you intend to treat.

If you stick to the guidelines we have discussed in this guide, and the advice you receive from a qualified dermatologist or doctor, you should be able to use red light safely and from it, derive the many tangible benefits we discussed in earlier parts of this book.

Do You Like My Book & Approach To Publishing?

If you like my writing and style and would love the ease of learning literally everything you can get your hands on from Fantonpublishers.com, I'd really need you to do me either of the following favors.

1: First, I'd Love It If You Leave a Review of This Book on Amazon.

2: Check Out Books

Visit my author profile on Amazon to check the latest books I have on red light therapy and related topics.

To get a list of all my other books, let me send you the list by requesting them below: http://bit.ly/2fantonpubnewbooks

3: Grab Some Freebies On Your Way Out; Giving Is Receiving, Right?

I gave you a complimentary book at the start of the book. If you are still interested, grab it here.

5 Pillar Life Transformation Checklist: http://bit.ly/2fantonfreebie

www.ingramcontent.com/pod-product-compliance
Lightning Source LLC
Chambersburg PA
CBHW020757220326
41597CB00012BA/568